The
Little Book
of
Snow

Sally Coulthard is a bestselling author whose titles include *The Little Book of Building Fires* and *The Hedgehog Handbook*. She also writes about rural life, craft and design. She lives in North Yorkshire on a smallholding at the bottom of a steep hill that's perfect for tobogganing.

The
Little Book
of
Snow

SALLY COULTHARD

An Anima Book

This book was first published in the UK in 2018 by Anima,
an imprint of Head of Zeus Ltd

9 7 5 3 1 2 4 6 8

A catalogue record for this book is available from

the British Library.

ISBN (HB): 9781788545792
ISBN (E): 9781788545808

Typeset by Lindsay Nash
Illustrations by Ian at KJA Artists

Printed and bound in Germany
by CPI Books GmbH, Leck

Head of Zeus Ltd
First Floor East
5–8 Hardwick Street
London EC1R 4RG

WWW.HEADOFZEUS .COM

'And when old Winter puts his blank face to the glass,
I shall close all my shutters, pull the curtains tight,
And build me stately palaces by candlelight.'

Charles Baudelaire, *Les Fleurs du Mal*

For James

CONTENTS

Out of the bosom of the Air,
Out of the cloud-folds of her garments shaken,
Over the woodlands brown and bare,
Over the harvest-fields forsaken,
Silent, and soft, and slow
Descends the snow.

'Snow-flakes',
Henry Wadsworth Longfellow

INTRODUCTION

My parents have a photograph of me as a child. I'm about eight years old in it and standing next to me is my best friend. We're in the doorway of our old house and, behind our heavy fringes and freckles, we're grinning from ear to ear. I can remember that picture being taken; we'd just found out that school had been cancelled, torn off our itchy uniforms and hurriedly dressed in the warmest clothes we could find. I'm wearing mismatched gloves, a tatty bobble hat and dungarees. My friend has grabbed a thin coat and her sister's wellies – she's holding two empty feed bags from her dad's farm. We look cold and impossibly cheerful. It's snowed heavily overnight. And we're going plastic bag sledging.

Waking up to a world covered in snow still has that effect, more than thirty years later. The view has changed – from suburb to open fields – but the feelings are still the same. That unexpected white blanket has an extraordinary power – the quality of light, the peaceful muffling, the way snow can cloak familiar objects into new shapes and sculptures. It's transformative.

For us, as kids, snow was synonymous with fun. Snow meant speed, exhilaration and toppling laughter. Snow gave us the freedom to fight, slide, crash and make a mess, without the fear of a sound telling off. Snow put a spanner in the works, stopped the numbing timetable of lessons and stole an extra day of free time from the working week. It didn't matter if you had an expensive wooden sledge or a black bin liner, snow was democratic – everyone was entitled to its pleasures.

As an adult, snow can be a mixed blessing. It still has the power to stop us in our tracks with its mesmerising beauty, but it can also mean traffic delays and the anxiety of a broken routine. Snow, when it really means business, can and has brought the country to a standstill, reminding us not to get too complacent. I like that though – we need a poke in the ribs once in a while, just to put us in our place.

The farm where I live with my young family nestles in the bottom of a valley. Access to the house is down a narrow, steep track, about a third of a mile (half a kilometre) long. When the snow comes, it turns the road into a giant frozen slide, lethal for the farm machinery and vehicles that need to come and go. Two particularly bad consecutive winters, about ten years ago, left us marooned for weeks – the snow was so deep we had to trudge to the top of the hill and flag down passing farmers to take us into town. We soon learned our lesson, scraping together enough money to buy a second-hand 4x4 that could cope with whatever the snow threw at us.

So, for my husband, winter can mean back-breaking dawns spent shovelling, scraping or blowing drifts to clear our only route out to civilisation. And yet, he's not averse to the lighter side of snow. A brilliant skier, for him snow represents a chance to break away from being responsible and reclaim the giddy pleasure of hurtling down a mountainside; for him, and many other people, skiing and other winter sports, are about total immersion – few pastimes allow such a cocktail of personal freedom, self-expression, adrenaline and cracking views.

We had another snowy winter this year. While we were killing time, watching the flakes tumble past the window, my youngest daughter – who's five – asked a flurry of questions, including 'Where does snow came from?' and 'Why is snow white?' I realised I didn't know. And so began the process of writing this book – it's a miscellany really, an assortment of things about snow, ice and winter weather that you might not already know. I wanted to find out answers to questions like 'Why is snow so *squeaky*?', 'What's the best snow for snowballs?' and 'Where's the coldest place you could choose to live?', and at the same time, try to get to grips with a little of the science of snow. Does anything grow in the Arctic, for example, or how do animals cope with freezing temperatures? And what about humans? How have we survived the ravages of cold weather for the hundreds of thousands of years we've inhabited our planet?

What's clear, from researching and writing this book, is that our climate is changing. The effects of this are particularly acute at the Polar Regions, but the ripples are being felt across the globe. Some of the changes can feel counterintuitive – how can global warming create *more* intense snow storms across the US and, at the same time, cause spring to arrive earlier than it used to? The answers are complex, and only touched on here, but it's important to raise the questions in the first place. We have many of the practical solutions at hand, but it takes political will and social pressure to address these issues effectively.

Because the reality is that we *need* snow; primarily because it is a vital cog in the environmental wheel and without it we are, for want of a better word, knackered, but also because it represents such a core feature of who we are, our shared culture and heritage. It might seem fatuous to fret about what Christmas would look like without snow, when we should be worrying about rising sea levels, but the two things go hand in hand. If we let ourselves imagine a world without snow, only then do we really start to understand what we might lose.

But let's not start on a glum note. This book is ultimately a celebration of snow – the science, the history, the relationship between us and the weather, and a brief exploration of how snow is intertwined with so many of our cultural references and celebrations. Oh, and it's also an insurance policy so that I have the answers ready the next time my snow-obsessed daughter asks a tricky question…

The hard soil and four months of snow make the inhabitants of the northern temperate zone wiser and abler than his fellow who enjoys the fixed smile of the tropics.

'Prudence', Ralph Waldo Emerson

SNOW
SCIENCE

WHAT IS SNOW?

There is water in the air, everywhere. You can't see it – it wafts about as vapour – but it's there. As this moist air is warmed by the earth, it floats upwards into the sky. If the sky is cool, this warm, watery vapour condenses into cloud droplets and falls back to earth as rain.

If the sky is freezing cold – 0°C (32°F) or less – something different happens. Clouds are mostly air and water but they also contain tiny specks of dust, pollen and other small particles. If a speck gets really cold, the water vapour sticks to it and freezes it into a tiny ice crystal. As the ice crystal gets bigger, it also gets heavier and begins to fall back to earth – this is snow.

Can it be too cold to snow?

Well, yes *and* no. The atmosphere must have moisture in it to create snow. Very cold air (-20°C/-4°F or less) is also usually very dry, making snowfall unlikely. In Antarctica's McMurdo Dry Valleys, for example, there is very little snow despite it being very cold (temperatures can be as low as -68°C/-90.4°F); a combination of low humidity and drying winds mean that, although the temperatures in the Dry Valleys are cold enough to snow, there just isn't enough water vapour in the air for it to happen.

In certain instances you can get very cold air that is also moist (over the sea, for example, or near products of combustion for heating). In these instances, the ice crystals usually stay suspended in the air. This is called 'ice fog'.

SNOWFLAKES

To appreciate the beauty of a snowflake
it is necessary to stand out in the cold.

Aristotle

If you could look through a microscope, at a tiny snowflake while it was forming, this is what you would see:

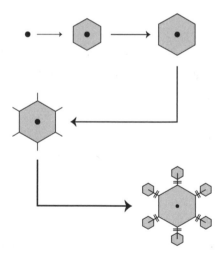

First, a speck of dust, floating around in the atmosphere. Water vapour in the air begins to stick to the speck, creating a droplet of water. The droplet then freezes into a tiny ball of ice. More water vapour sticks to the ball of ice, making it grow into a six-sided ice crystal. As the crystal grows, the six corners start to branch into a star shape. As the snowflake tumbles to earth, the branches keep growing, sprouting smaller branches until you have a fully formed snow crystal.

Are two snowflakes ever alike?

Snowflakes get their unique shape as they fall from the sky. Changes in humidity and temperature affect their shape – the more moist and warm the air, the more complicated and beautiful the patterns become. As the snowflake journeys from its cloud to the ground, twirling and swirling, it will fall through different atmospheric conditions. Changes in the surrounding temperature and humidity cause the snowflake to grow in different directions – one minute it might be extending its six long arms, the next it could be creating side branches or filling in between the gaps. Although snowflakes always have their essential six-sided shape,[1] the variations are almost endless – no two snowflakes fall and develop in exactly the same way. Snowflakes can also change shape if they bump into each other, clumping together to form larger snowflakes.

Scientists estimate that around 24,000,000,00 0,000,000,000,000,000 (24 septillion) snowflakes fall to earth every year and that the possible number of different shapes exceeds the number of atoms in the known universe – i.e. it's very unlikely to find two snowflakes that look alike. If the air is very cold, snowflakes can stay as simple hexagons – called plate crystals – and these can

1 You can get snowflakes with twelve points – these are called 'twin crystals' and happen when two crystals grow on top of each other, onto the same speck.

look very alike under a microscope. However, on a molecular level, even these small snow crystals are unique.

SLEET and HAIL

Sleet starts off as rain but freezes into ice pellets if it falls through a very cold layer of air between the clouds and the ground. Hailstones, on the other hand, are formed in warm thunderstorms; raindrops that have formed at the bottom of a cloud are driven upwards, into the top of the cloud, where it's much colder. The raindrops freeze and fall back to the bottom of the cloud, only to be carried upwards on another updraught and refrozen. The hailstone will only fall to the ground once it's heavy enough, or the updraught has weakened. If you slice through a hailstone it has rings, like a tree trunk, which will tell you how many times it was carried to the top of the cloud.

◇◇

Did you know? The biggest hailstone ever recorded fell during the summer of 2010 onto the town of Vivian, South Dakota. The hailstone measured 20.5 centimetres (8 in) wide and weighed almost 1 kilograme (2.2 lbs). The largest hailstone ever to land in the UK plummeted onto Horsham, Sussex back in 1958, with a 6.35 centimetres (2½ in) diameter and weighing 190 grams (7 oz). While the UK hailstone weighed slightly more than a cricket ball, the US record would have been the equivalent of a bag of sugar dropping from the sky.

◇◇

Megacryometeors

The biggest piece of ice ever to fall from the sky, however, wasn't a hailstone, or at least not as we know it. In 1849 Scotland's *Ross-shire Advertiser* reported the following incident:

> 'A curious phenomenon occurred at the farm of Balvullich... on the evening of Monday last. Immediately after one of the loudest thunder peals ever heard there, a large and irregularly shaped mass of ice, reckoned to be nearly 20 feet [6 m] in circumference, and of a proportionate thickness, fell near the farmhouse. It had a beautiful crystalline appearance, being nearly transparent, if we except a small portion of it, which consisted of hailstones of uncommon size fixed together.'

There's disagreement in the scientific community about what this ice chunk, and the many others that have fallen across the globe in the intervening years, really are. Research into what have been dubbed 'megacryometeors' suggests that these vast lumps are ice are neither formed in outer space nor the result of aircraft leaking water at height. Their composition is similar to hail, so one theory is that they may be caused by specific atmospheric conditions triggered, at least in part, by global warming. Just as when a hailstone is formed, megacryometeors may be created when an ice crystal is forced, repeatedly, through moist air and then refrozen, acquiring layer after layer of ice. Climate change has increased both turbulence and water vapour levels in the earth's atmosphere; if current weather patterns continue megacryometeors may become even more common in the future.

DIFFERENT TYPES of SNOW

'Hear! hear!' screamed the jay from a neighboring tree, where I had heard a tittering for some time, 'winter has a concentrated and nutty kernel, if you know where to look for it.

Henry David Thoreau

Wet snow

If snow falls through moist air that is slightly above freezing – but less than 2°C (35.6°F) – the snow crystals stick together and make larger, fluffy flakes. This is called 'wet snow' and, because it's sticky, is perfect for snowball fights and building snowmen (see **SNOW PLAY**, page 131). Wet snow can be difficult, however, because it's heavier than dry snow. This makes it hard to shovel and causes problems when it weighs down power cables, telephone lines and tree branches.

Dry snow

If the snow falls through really cold, dry air (0°C/32°F or less) the crystals remain small and powdery. This is called 'dry snow' and is ideal for snow sports such as skiing and sledging. Dry snow is easier to clear away than wet snow. Because it's light, however, it also has a tendency to get whipped up by the wind and create snow drifts.

How much water does snow contain?

There's an old wives' tale that '10 inches of snow equals 1 inch of rain'. But how true is it? Meteorologists look at something called the 'Snow Ratio'. This is how much snow you get for a given amount of water. Say you have a snow ratio of 10:1 – that means if you melted 10 inches of snow you'd be left with 1 inch of water. Dry, powdery snow has a high ratio – usually around 15:1 or 20:1, while heavy, wet snow can be as low as 5:1. The average ratio of snow is actually around 10:1, so it turns out the old cliché is true.

SNOWFALL

About a million billion kilograms of snow falls every year and depending on other conditions – such as wind speed or temperature – it creates a wide range of weather conditions. Meteorologists classify some of these into groups, each with its own set of characteristics:

BLIZZARD The term is often bandied about, but weather forecasters only officially use the term 'blizzard' when three conditions are met: a storm needs to have large amounts of snow or blowing snow with winds in excess of 35 miles per hour (56 kph); there needs to be visibility of less than a quarter of a mile (0.4 km) *and* the storm needs to last for at least three hours. The word 'blizzard' used to mean a violent blow or volley of ammunition – the first example of its use as a weather term was probably in the US, around the 1860s; early German settlers would describe severe winter storms as coming *Blitzartig,* meaning 'like lightning'.

SHOWERS, FLURRIES and SQUALLS The difference between these three terms is usually the duration, amount of snowfall and how much snow remains on the ground. They're fluid terms, and open to interpretation, but as a general rule: a snow **SHOWER** has a short duration, with moderate snowfall, and some accumulation on the ground; a **FLURRY** is a light, intermittent snowfall, which only lasts a brief time and doesn't tend to settle; and a **SQUALL** is an intense, but short-lived, period of moderate-to-heavy snowfall, accompanied by strong winds, where snow also tends to accumulate on the ground.

LAKE-EFFECT SNOW This is a type of snow squall when cold, dry winds blow across large lakes of warm water, picking up water vapour, which then freezes and falls as snow when it reaches land. Lake-effect snow squalls tend to be intense, dumping huge amounts of snow in just a few hours; one of the most famous storms hit upstate New York in 2006, when an unusually cold air mass swept across the Great Lakes – Cowlesville, near Buffalo, had 2.3 metres (88 in) of snow in just one week. 'Sea-effect snow' is the ocean equivalent.

GRAUPEL (also known as snow pellets or soft hail) When a snowflake tumbles from the sky, it can travel through lots of different atmospheric conditions, from warm moist air to freezing cold layers. Graupel is the name for small pellets of ice that are created when supercooled water droplets freeze themselves onto a snowflake as it falls. Snowflakes that have this frozen coating are often referred to as 'RIMED' and are often so thickly covered that you can't see their original snowflake shape. Unlike hail, which forms in warm thunderstorms, however, graupel falls in winter. And, while hail is hard, graupel is soft and falls apart when you touch it. The word 'graupel' is originally German and derives from *Graupe*, meaning 'grain of cereal'.

SNOW COVER

Once snow has settled on the ground it will do one of two things: it'll either disappear or it'll stay put. Let's deal with the second of these two scenarios first:

In very cold climates, snow can stay on the ground for years on end – and is known as **PERENNIAL SNOW**. The longer it stays on the ground, the more densely packed it becomes, eventually turning into glacial ice.

Depending on how old the snow is, it's known

by different names: young snow, which has fallen, partially melted and then refrozen is called NÉVÉ. If this same snow survives for a whole year, it becomes FIRN – a dense, granular halfway-house between snow and ice. Subsequent snowfalls will continue to add to this layer of FIRN, gradually weighing it down and squashing out any air pockets until it becomes dense, GLACIAL ICE.

But snow can also just last for a short while – this is SEASONAL SNOW. Very fresh, recently fallen snow – where you can still see the original shape of the snowflake crystals – is called NEW SNOW. One type of new snow, which has a light, loose texture and a low water content, is called POWDER.

If the surface of powder snow starts to melt and then refreezes (which can be caused by sunlight, rain or wind) you get a layer of thin CRUST. Crusts can be described as 'breakable' or 'unbreakable' depending on whether they'll support the weight of a turning skier; or 'zipper' which means a skier can break the crust and ski through it.

In fact, skiers have a wonderfully rich and varied vocabulary to describe the almost infinite snow conditions they come across; from the much sought after 'champagne powder' (light, fluffy snow with very low moisture content) to mashed potato (mushy, slow-going spring snow), corduroy (freshly groomed snow with thin, parallel grooves) to snorkel (airborne snow that makes it difficult to breathe).

FROST and HOAR FROST

The cold hoar-frost glistened on the tombstones, and sparkled like rows of gems, among the stone carvings of the old church.

The Posthumous Papers of the Pickwick Club, Charles Dickens

Sometimes frost can be so dramatic that it can look like snow. This is called 'hoar frost'. Normal frost – also called 'ground frost' – forms when water vapour in the air condenses on solid surfaces and then freezes as the temperature drops to below 0°C (32°F). Hoar frost happens when water vapour comes into contact with surfaces that are *already* below freezing – the ice crystals start to grow straight away, and continue to get bigger as they 'grab' more water vapour from the air. Hoar frost has a feathery, hair-like appearance, which is where it gets its name from: *hoar* is an Old English word meaning 'white-haired with age'.

Sublimation

We've all seen snow melt, but did you know it can also turn directly from a solid back into a gas – a process similar to evaporation, called 'sublimation'? Under a certain set of weather conditions – low temperatures, strong winds, intense sunlight – snow on the ground will vaporise back into the air before it has had chance to turn into slush.

SNOW FORMATIONS

Snow is a dynamic, moving entity. As it lies on the ground, snow is subjected to weather conditions that not only change its composition but physically change its shape, sometimes with startling results:

SNOW ROLLER – this rare event creates fat tubes of snow that look like rolls of loft insulation. Just as a small snowball will get bigger if you roll it through loose snow (see **MAKING A SNOWBALL**, page 141), so a ball of snow can be blown along the ground and pick up more snow as it goes. A snow roller can get as large as 1 metre (3 ft) in diameter. They tend to be hollow, however, so the centres often collapse in on themselves.

CORNICE – from the Italian for 'ledge', a cornice is a mass of overhanging snow. These are formed when wind whips up snow and drops it on the steep leeward side of a mountain top or ridge (the leeward side is the opposite one from where the wind is blowing). This can create a huge curl of snow which extends out over thin air. And, while they're beautiful to look at from below or the side, they're often invisible from above, creating a lethal trap for any walkers who happen to venture on to one by mistake.

BRIDGES – if a cornice grows big enough to span the gap between two ridges, or create an arc across a crevasse, it becomes a snow bridge. These deadly formations can create the illusion of a solid surface, luring climbers to cross them only to find they are walking on a thin layer of snow over fresh air. And, because they're thicker and stronger at the point at which the snow bridge meets the edge, climbers often don't discover the fact until they're at a point of no return.

PENITENT SNOW – penitents look like fields of church spires or people kneeling, penitently, in prayer. They're columns of old compact snow or glacial ice, wider at the base than at the top, and can reach as high as 6 metres (6½ yards) tall. They're formed by prolonged action of the sun in a dry, cold atmosphere. Charles Darwin was the first person to record these extraordinary formations in 1835, when he noted a grim discovery – a frozen horse impaled on a penitent, *'with its hind legs straight up in the air, the animal must have fallen into a hole head downmost and thus have died'. Beagle diary, 22nd March 1835*

BARCHANS – Snow, just like sand, can be easily whipped, carried and dropped by the wind to create small, wave-like ripples, snow dunes and barchans. A barchan is a snow dune that looks like a horseshoe, with its 'horns' pointing downwind. Once the wind has died down, if a dune or barchan then refreezes and consolidates, further gusts of wind can erode its shape, sculpting it into jagged, frozen waves called SASTRUGI.

GLACIERS

During the Ice Age, glaciers covered about a third of the Earth's land surface; today, this figure is nearer a tenth, but still represents a vast area. Nearly all the remaining glacial ice sits in just two places – Greenland and Antarctica – and these enormous glaciers, called ICE SHEETS, hold three-quarters of the world's fresh water. Glaciers currently cover about 6 million square miles (15 million km^2) – if all this ice melted overnight it would raise sea levels by about 70 metres (230 ft), enough to submerge half of the UK, including London.

Glaciers also 'flow' under their own weight, like very slow rivers, carving out crevasses and dragging huge chunks of rock as they go. Some barely move, while others zip along at speeds of up to 30 metres (98 ft) per day. Occasionally, glaciers can also experience 'surges', periods of rapid movement caused by melting water which lubricates and releases the glacier from its position. The Kutiah Lungma Glacier in Pakistan holds the current record for the fastest surge – in 1953 it moved a terrifying 7½ miles (12 km) in just three months.

For a glacier to form, it needs to be cold enough for snow to remain all year round. Every year's snowfall compresses the last, compacting the layers and, eventually, creating glacial ice. (Some glaciers are spectacularly old – the Antarctic ice sheets are thought to be at least 40,000,000 years old.) Snow that falls on glaciers each year captures many of the things that are floating around in the atmosphere at the time, including dust, ash and human pollutants. By taking samples of these glaciers (see ICE CORES, below) we can track changes in climate and environmental conditions.

ICE CORES

Ice cores are cylinders drilled from glaciers. They're useful because they can tell us information about changes in the environment over thousands of years, especially temperature fluctuations and greenhouse gases.

Most of the ice core records we have come from the two large ice sheets over Antarctica and Greenland and cover a remarkably long period of time. The oldest ice core – drilled out of the Antarctic plateau – reveals information going back a staggering 800,000 years. By measuring the ratios of different water isotopes, scientists are able to 'read' how the world has experienced a succession of long, frozen glacial periods, broken up every 100,000 years or so by warmer interludes. The ice cores also show us that the last Ice Age ended about 11,000 years ago and we're currently enjoying one of the balmier, interglacial periods.

Ice cores also tell us about the damage we're

doing to the planet. Within the ice, small bubbles of trapped air contain snapshots of the health of the atmosphere at different times of history. The results show that concentrations of two greenhouse gases – carbon dioxide and methane – have rocketed since the Industrial Revolution and that the rate of increase is unprecedented over the last eight hundred millennia. We also know that during the 20th century the Earth's temperature rose by 2°F – it doesn't sound much but tiny fluctuations can translate into huge ecological changes. At the end of the last Ice Age, for example, when much of the US and Europe was under large ice sheets, average temperatures were only around 5°-9°F cooler than now (also see **SNOW** and **CLIMATE CHANGE** page 62)

ICICLES

For icicles to form you need three things: first, you need a source of frozen water (say a roof covered with snow or a gutter filled with ice); second, you need a blast of sunshine to begin melting the frozen water so it starts to drip; and third, you need sub-zero air temperatures so that the dripping water begins to refreeze.

The first droplet of water to refreeze creates the beginning of an icicle. As more water drips down the sides of this new icicle it also refreezes, making it grow. But that doesn't explain why icicles are carrot-shaped, tapering to a narrow tip at the bottom. If the water froze uniformly, an icicle should be ball-shaped, growing bigger with every new layer of frozen water.

What happens is extraordinary. An icicle is actually warmer at the top than at its pointy tip. As water refreezes around the icicle, it releases heat into its immediate surroundings. This slightly warm air rises, making the top of the icicle warmer than its tip. As more water runs down the surface of the icicle in a very thin film, some of the water freezes around the sides but most freezes at the tip. So, icicles grow longer quicker than they grow fatter.

◇◇

Did you know? The ancient church of St Michael and All Angels, in Bampton, Devon, displays one of the most unusual epitaphs ever written. An eighteenth-century memorial plaque fixed to the tower tells the unfortunate story of a young boy, who was killed by a falling icicle:

IN MEMORY OF THE CLERK'S SON

Bless my i.i.i.i.i.i. [eyes]

Here he lies

In a sad Pickle

Kill'd by Icicle

IN THE YEAR 1776.

◇◇

Scientists have also recently worked out why icicles are usually rippled. It's due to impurities in the water, such as salts. When researchers grew icicles in the laboratory using distilled water, they were completely smooth. They also discovered that, contrary to expectations, icicles grew into strange, forked shapes when the surrounding air was still. When the air was constantly moving, the icicles grew perfectly carrot shaped. Quite why they do this, we still don't know.

THE COLOUR OF SNOW

Let's start with how you actually *see* colour. Light travels in waves and consists of a rainbow of colours – when that light hits an object, some of the colour waves bounce back and some are absorbed by the object. A red apple, for example, is red because the apple bounces back the red light but absorbs all the other colours.

When light hits snow, *all* of the colours are

bounced back in equal amounts, so all we see is white. This is because all the fluffy, freshly fallen snowflakes contain lots of tiny air bubbles and crystal edges that refract and reflect the light.

But have you noticed that sometimes snow can look blue, especially as it becomes compacted into ice? As snow is compressed into ice, the small ice crystals merge together into larger ones, and many of the air pockets are squeezed out. This allows the light to penetrate the ice, where some of the colours get absorbed. Both water and ice absorb red and yellow wavelengths but reflect back blue and green. That's what gives icebergs their distinct colour.

Pink snow

Pink snow or 'watermelon snow' is found in both polar and alpine regions. It was originally thought to be caused by mineral deposits but more recently scientists have discovered that the culprits are cold-tolerant algae called

Chlamydomonas nivalis. The algae contain a strong red pigment which is designed to absorb heat, melting the snow to provide the algae with a fresh water supply.

SNOW and SOUND

'Tis winter, yet there is no sound
Along the air,
Of winds upon their battle-ground,
But gently there1
The snow is falling, —all around
How fair – how fair!

'Snow', Rev. Ralph Hoyt

Have you ever noticed how quiet everything goes when it snows? It's true that most animals and people stay indoors or hunker down when it's cold – so there's less noise being made – but there's also a scientific explanation for this muffling effect.

Snow, it turns out, is pretty good at absorbing

sound. Hard, shiny surfaces such as glass and metal reflect sound, bouncing it around – think of the acoustics in a large building, for example. Porous materials, on the other hand, such as sponges or foam pads, absorb sound waves, dampening down their effects. A thick layer of freshly fallen snow is just like a sponge, full of hollow pockets of air that capture the sound waves and deaden them.

In fact, a couple of inches of snow behave a lot like commercial sound-absorbing materials; these materials are measured on a scale of 0 to 1 (0 means that a material absorbs no sound, 1 means a material absorbs all sound. The higher the number, the better the material is at soaking up sound.) Snow is roughly 0.6, which means it absorbs 60 per cent of the sound waves that hit it. Carpet, by comparison, only absorbs about 0.1 or 10 per cent.

Why does snow squeak when you walk on it?
When you walk on snow, your foot compresses the ice crystals. When the ice crystals rub against each other they can create friction, which in turn can create a squeaking or crunching noise. The colder the temperature of the snow, the greater the friction between the ice crystals. At warmer temperatures, the ice crystals slide against each other, making little or no noise. The crucial temperature at which snow starts to crunch seems to be about –10°C (14°F); any warmer than that and it stays quiet.

FAKE SNOW

Some of the most memorable films or TV shows have been set against a backdrop of snow or ice. From *Dr Zhivago* to *It's a Wonderful Life,* many a special effects team has had to come up with ingenious ways to transform an unpromising, often sweltering Hollywood set into a dramatic winter scene. Early attempts often made life difficult for both cast and crew members – in the 1930s and '40s, for example, white asbestos was marketed under brand names such as 'White Magic' or 'Snow Drift' and famously heaped onto the set of *The Wizard of Oz*. Other movie cheats have included soap flakes, cornflakes painted white, polystyrene, urea formaldehyde, marble dust and, in the case of the Superman movies, vast quantities of salt.

It's ironic that re-creating one of nature's purest, most beautiful phenomena involved such toxic ingredients. Thankfully, today's movies use blown, recycled paper – a technique so versatile it can be sprayed onto sets, made into snowballs and even leaves footprints and tyre marks.

But paper snow won't get you very far if you need a mountainside's worth; ski resorts are increasingly finding that they need to supplement the winter season's snowfall – the European Environment Agency estimates that the average length of the snow season has shortened by five days every decade since the 1970s. That leaves many ski resorts – especially those at low altitudes – with no option but to use snow-machines, which force water and air through cannons at high pressure to create 'instant' snowfall.

Ecologically this is problematic; not only is 'artificial snow' hugely costly in terms of energy and water use (it takes over 380 litres/100 gallons of water to make just 1 cubic metre/35 cubic ft of snow) but, because the water comes from mountain reservoirs, the snow contains more nutrients and minerals than real snow. When it melts, this nutritionally over-rich 'artificial snow' changes the balance of the mountainside flora, interferes with the water table and melts later than 'real snow', postponing the normal flow of water into the valleys.

Though a tremor of the winter
Did shivering through them run;
Yet they lifted up their foreheads
To greet the vernal sun.

'The Crocuses', Frances Ellen Watkins Harper

SNOW
ECOLOGY

WHERE DOES IT SNOW?

It's always snowing *somewhere* in the world. Polar regions see year-round freezing conditions, but there are also plenty of high-altitude places – such as the Andes and the Rocky Mountains – where it's also cold enough to get snow 365 days of the year.

Generally speaking, the 'north' gets more snow cover than the 'south'. That's because nearly 70 per cent of the Earth's landmass sits in the northern hemisphere, a good chunk of it within the Arctic Circle. During the month of January, scientists estimate that at least half of the entire northern hemisphere is covered with snow, especially Russia, Northern Europe, Canada, parts of Siberia and Alaska.

The northwest side of Japan also gets huge amounts of snow, thanks to the 'sea-effect' (see **LAKE-EFFECT SNOW** page 25) – icy-cold winds blow across Siberia, pick up moisture from the Sea of Japan, and then dump their contents as snowfall over the Japanese Alps. It's been estimated that the snowiest area of this region – that sits in the Nagano Prefecture – may get as much as 38 metres (125 ft) of snow each year.

In the Southern Hemisphere (apart from the Antarctic), snow tends to accumulate only in the mountainous regions of places such as New Zealand and South America. Even countries near the warm equator, however, can get snow – Mount Kilimanjaro, for example, is high enough to get year-round ice and snow on its upper reaches.

The coldest places on Earth

On a high ridge on the East Antarctic Plateau, scientists used satellites to record the lowest temperature ever at –92°C (–133.6°F). The place is, unsurprisingly, uninhabited. Any human finding himself there by mistake would be dead within three minutes.

The coldest permanently inhabited place on earth is the small village of Oymyakon, in the Siberian tundra. In 1933, a temperature of –67.7°C (–89.9°F) was recorded at Oymyakon's weather station.

WHY DO WE NEED SNOW?

When your car wheels are spinning in yet another snow drift, the question of why we need snow is probably the last thing on your mind. And yet snow and ice are absolutely central to the health of the planet and the functioning of its ecosystem.

One of snow's most important jobs is to act as nature's sunscreen. Snow and ice have the perfect reflective surface – like a vast, white linen sheet – bouncing back 80–90 per cent of the sun's radiation and keeping the Earth deliciously cool. Snow cover regulates not only the temperature of the planet but also affects regional weather patterns, such as the duration of summer rain.

Both snow and ice also act as a storage mechanism for much of the world's supply of fresh water, releasing it slowly into rivers, soil and reservoirs and providing a steady, reliable supply for people, plants and animals. More than one-sixth of the world's population relies on the slow melt of seasonal snow for their water supply. In the western United States, for example, nearly three-quarters of the annual streamflow that provides the water supply comes from melting snowpacks in the mountains.

Permanently frozen ground – called permafrost – has the important task of slowing down the decomposition of any organic matter that it contains: if it were to melt and heat up, the rate of decay would increase, releasing more carbon dioxide and methane, both of which would contribute to further climate change.

Conversely, in places where the ground isn't frozen but is covered in a layer of snow, the snow acts as a thermal blanket, protecting the soil and microorganisms from extreme changes in air temperature.

◇◇

Did you know? Syracuse, New York, once banned snow. On 30 March 1992, after the city had been battered by record snowfall – over 4 metres (162 in) – the Syracuse Common Council unanimously approved the resolution: 'Be it resolved, on behalf of the snow-weary citizens of the city of Syracuse, any further snowfall is expressly outlawed in the city of Syracuse until December 24, 1992.' The resolution didn't make a jot of difference, and was never designed to, but it certainly helped the city smile through a particularly snowy season.

◇◇

SNOW and PLANTS

O Wind, if Winter comes, can
Spring be far behind?

'Ode to the West Wind', Percy Bysshe Shelley

When snow covers the garden, bringing everything to a standstill, it can be tempting to wish for warmer winters. And yet the coldest season of the year – with its snowfall, ice and biting winds – is an essential part of the growing cycle for many plants.

As we've already learned, a layer of snow creates an excellent insulating blanket against the worst of the winter weather and helps to release moisture back into the soil as it slowly melts. But what about ice-cold temperatures and frozen soil – do plants actively need winter to thrive?

When we groan at the damage done by a late frost or heavy snowfall, it's tempting to think that warmer winters will be kinder to plants, encouraging more growth and productivity. And yet, for the most part, long periods of cold weather are hugely beneficial for the plant world.

Many plants simply won't flower or produce fruit without a spell of cold weather; this is thanks to a process called 'vernalisation', where plants go into a period of dormancy during cold temperatures to help them prepare for the following spring. Apple and peach trees, for example, need a minimum amount of exposure to cold weather if they're going to successfully fruit the next year (a period called 'chilling time').

Ask any gardener and they'll tell you that a cleansing, cold winter is often a blessing. Many of the fungal problems that plague gardens during a warm, wet summer are wiped out by a decent spell of freezing temperatures, as are other pests such as overwintering aphids. Heavy clay soils or clods of earth are broken down by hard frosts, leaving behind a more workable tilth, while cold snaps transform starches into sugars in crops such as parsnips, making them even more delicious.

In the polar regions, where much of the ground is permanently frozen, plants have to deal with year-round winter conditions; amazingly, many thrive in such harsh conditions and have adapted to take advantage of the local terrain. Around 1,700 species live on the Arctic tundra alone, including flowering plants, shrubs, grasses and herbs. Only a very thin layer of soil thaws every year – plants keep themselves compact, close to the ground and shallow-rooted to compensate; even the 'trees' are tiny – the *Salix arctica*, or Arctic willow, grows no higher than 9 centimetres (3½ in) tall, for example, spreading out in clumps like a miniature forest. Leaves, stems and seeds are covered in fuzz, to protect against the icy winds, while many Arctic species can grow even under a layer of snow or at extreme temperatures. For the few days that there is a polar summer, plants have learned to develop quickly – making the most of the intense, but short-lived sunlight. The Arctic poppy, for instance, turns its custard-yellow head to follow the sun, the cup-shaped petals helping to focus the sun's rays.

SNOW and ANIMALS

Just as plants have adapted to cope with the snow, so too have animals. For some, the trick is to hunker down and ride out the storm. For others, the coming of winter signals the start of a mass migration to warmer climes.

From birds to butterflies, many species of animals choose to move when the temperatures drop. Sometimes the distances are modest – a frog, for example, may hop from its shallow summer breeding pond to a deeper lake nearby (which is less likely to freeze), while other creatures travel astonishing distances – the Monarch butterfly flutters from Canada and Mexico in search of winter warmth, for instance, while the Arctic tern makes a round trip of 44,000 miles (70,810 km) from pole to pole *each* year – over a lifetime that adds up to 1.25 million miles (2,012,000 km) or three trips to the Moon and back again.

For other creatures, the strategy is simple – stay put and survive. Some of the smaller mammals, reptiles and invertebrates hide under the thick blanket of snow, making the most of its insulating protection; others, such as bats and dormice, hibernate – effectively going into a period of suspended animation to wait out the season. European hedgehogs, for example, hibernate between December and March, depending on the weather. They do this not just to cope with extremes in temperature, but also because insects – their main food source – disappear for the winter.

For other creatures, who don't hibernate but find their food sources greatly depleted, stockpiling is the only option. Squirrels aren't the only animals to hide pockets of food; moles hoard earthworms in mounds of soil, foxes will bury prey to last out the winter, and mice tuck away stores of seeds and nuts in their nests. The hamster-like American pika – an animal that's related to rabbits and lives in rocky terrain – copes with winter by eating dried vegetation, such as grasses, that it made into little hay piles during summer.

And, of course, animals have developed physical adaptations that can endure otherwise perilous drops in temperature. Arctic hares, reindeer, polar bears, musk ox, wolves, even lemmings have coats of thick, insulating hair that keep them warm when temperatures drop below freezing. Mammals that live in ice-cold water survive thanks to a generous layer of fat tissue (called blubber), while penguins also huddle together in groups to keep warm and away from the wind. Perhaps one of the most extraordinary cold-weather adaptations belongs to the Alaskan frog. Thanks to a cocktail of chemicals in its body tissue (including urea and glucose) the Alaskan frog can spend seven months of the year frozen – like an ice lolly – with two-thirds of its body water as ice. Physiologically, it's almost dead – no heart beat, no blood flow, limbs stiff – and yet, come springtime, it simply thaws out and hops away.

SNOW and CLIMATE CHANGE

The past few winters have seen some heavy falls of snow in the US and the UK. For people who don't believe in the reality of climate change, the question has been: 'How can the world be warming up when we've got so much snow?' It seems counterintuitive.

The Earth's temperature is unequivocally rising. But that doesn't mean that all snow will disappear. Patterns of snowfall around the globe depend on a number of interconnected factors but, put simply, if temperatures rise, more water will evaporate from the sea into the atmosphere.

More moisture in the air means more rain and – if it's cold enough – snow.

But climate change also causes changes in the air and ocean currents, so this extra rain and snow won't fall evenly across the world. Scientists predict that while a warmer planet is likely to shorten the snow season and reduce the amount of snow that falls overall, when the conditions are cold enough for snow, the warmer, wetter air will create bigger, more dangerous snowstorms.

Extreme snow events

It's difficult to predict exactly what the effect of climate change will be on extreme snow events, but it's interesting to look at the data. Scientists in the US have been recording the frequency of extreme snowstorms over the past century and watched a gradual but accelerating rise in numbers. Statistics show that, for example, compared to the first half of the 20th century, between 1950 and 2000 the number of extreme snow storms in the US *doubled*; one theory is that the melting sea ice in the warming Arctic weakens the jet stream, allowing icy polar air to travel further south, hitting the eastern United States with dramatically cold winters and record snowfall.

◇◇

Did you know? The word 'Arctic' comes from the Greek word for bear, *'arktos'*. Some constellations are visible in the night sky all year round. These groups of stars are known as 'circumpolar', the most famous of the northern constellations being Ursa Major ('The Great Bear') and Ursa Minor ('The Little Bear'), hence the name. Antarctic means 'opposite the Arctic'.

◇◇

SNOW and SPACE

Recently, NASA scientists found evidence of what they think is water ice in the dark craters near the North and South Poles of the Moon – but does it ever really snow in space?

On Mars, where the average temperature is around –60°C (–76°F), it's certainly cold enough to snow and in 2008 scientists recorded water-based snow – just like our own – falling near Mars' North Pole. The snow doesn't seem to settle on the ground, however, thanks to Mars' thin atmosphere. Ice crystals fall too slowly through the atmosphere and evaporate before they have chance to accumulate into any kind of ground cover. Mars' South Pole, on the other hand, is covered with a layer of frozen carbon dioxide – basically 'dry ice' – which also falls from the atmosphere as CO_2 snow.

On Jupiter, scientists think that the clouds

hold a mixture of frozen ice and ammonia, which falls in the form of something between snow and hail, while one of Jupiter's moons – Io – has snowflakes made from sulphur. Other moons and planets have their own, curious versions of 'snow': Saturn's moon, Enceladus, has hot geysers which spew water into the atmosphere that then freeze and fall as snow; Pluto enjoys mountains capped with methane snow; and Triton, Neptune's largest moon, is frosted with nitrogen. But perhaps the oddest object in space has to be Kepler-13Ab – a vast, hot planet over 1,700 light years from Earth – there, it snows titanium dioxide, an ingredient that, ironically, is used commonly in skiers' sunblock.

How like a winter hath my absence been
From thee, the pleasure of the fleeting year!
What freezings have I felt, what dark days seen!
What old December's bareness everywhere!

'Sonnet 97', William Shakespeare

PEOPLE AND SNOW

THE ICE AGE

For millions of years, the Earth's climate has wobbled – alternating between warm, balmy intervals and harsh, frozen periods when snow and ice sheets clung to the continents.

The most recent of these Ice Ages began 2.6 million years ago and has seen glaciers advance and retreat over great swathes of the planet, often swallowing up much of North America, Europe and Asia. Within this long frozen spell, however, there have been many periods of respite; short, warm periods called 'interglacials', which last thousands of years only to fade back into another chilly phase. We're living in an interglacial period right now – one that started about 11,700 years ago.

Humans have been shaped by these periods of wildly changing climate and, crucially, their success has been due, at least in part, to their ability to adapt and be versatile. Huge leaps in human development – such as the discovery of fire or migration out of Africa – may have their origins in dealing with an ever-changing environment.

When you look at the archaeological record, we see a pattern of early humans battling against the snow and cold, sometimes winning the fight, sometimes being driven back to warmer climes. Take Britain, for example. At least four different species of humans have attempted to make the UK their home. An ancient set of footprints and the discovery of stone tools paint a picture of a very early human – *Homo antecessor* – attempting to eke out an existence 900,000 years ago. Back then, Britain was joined to France and would have been colder than today, with bitterly harsh winters, and few edible plants. To survive, these early pioneers would have scavenged, and maybe hunted, the mammoths, elk and wild horses that roamed the southeast.

Another species followed 400,000 years later; *Homo heidelbergensis*, but not for long; 450,000 years ago the climate took a nose-dive. Of all the snowy, glacial periods early humans experienced, this was the worst – for thousands of years Britain became uninhabitable.

About 400,000 years ago, it was the Neanderthals' turn to venture north, into Britain, once again crossing the land bridge that appeared when temperatures and sea levels dropped. Between then and 50,000 years ago, our resilient, capable cousins learned to exploit the land's resources – following rhino, deer and mammoth – and coping with extremes of cold. For a species who had no real shelter and may not have had access to fire making, scientists speculate that Neanderthals would have developed physiologically to cope with snow and ice, as well as inventing strategies such as storing food over winter.

Modern humans finally came to Britain around 40,000 years ago but, it seems, didn't stay for long – perhaps finding the northern temperatures just too cold. Only after the last gasp of the glacial period and temperatures finally began to rise – around 12,000 years ago – do modern humans finally feel able to make Britain their permanent home.

THE LITTLE ICE AGE

What is the world, O soldiers?
It is I:
I, this incessant snow,
This northern sky;
Soldiers, this solitude
Through which we go
Is I.

'Napoleon', Walter de la Mare

Have you ever seen paintings of the River Thames, with people whizzing about on ice skates, and wondered why that never happens these days? Although historians argue over the precise dates, between the mid 1300s and 1800s, there seems to have been a period of regionally cold conditions that's been dubbed 'The Little Ice Age'. So much of medieval history can seem grim – with reports of plagues, famines and failed harvests – and there's a certain amount of evidence to suggest that the weather and, in particular, the winters may have, indeed, been worse in the past.

While the overall drop in temperatures globally was probably only around 1°C (30°F) throughout this period, there were certainly parts of Europe and North America that experienced more pronounced dips, especially in winter. The effect was dramatic – it wasn't uncommon for the River Thames to freeze over, often for months at a time. The first River Thames Frost Fair was held in 1607 – a festival of ice skating, pop-up shops, debauched drinking and racing – and between then and 1814, Londoners enjoyed countless more. Canals and rivers froze, Alpine villages were swallowed by advancing glaciers, and farmers across Europe struggled with hugely variable growing seasons. Even the practice of witch hunting has been blamed on the Little Ice Age, with panicked peasants believing that witches were responsible for the terrible weather. One German chronicler, writing in the 1620s, noted:

> all the vineyards were totally destroyed by frost... the same with the precious grain which had already flourished... Everything froze, [something] which had not happened as long as one could remember, causing a big rise in price... As a result, pleading and begging began among the peasants, [who] questioned why the authorities continued to tolerate the witches' and sorcerers' destruction of the crops. Thus the prince-bishop

punished these crimes, and the persecution
began in this year...

On a happier note, there's even a suggestion that the Little Ice Age was responsible for the famous tone of Stradivarius' violins. Researchers have suggested that the cool climate of the Little Ice Age affected the growth rate of the timber used in his violins. Dense, slow-growth wood may have improved the acoustic quality of the violins, although it's still just a theory. What we do know, however, is that during the Little Ice Age, people adapted to the cold in different, often inventive ways – from the adoption of new, hardier crops to more efficient fireplaces, the increasing use of coal to warmer, thicker clothing – it seems few aspects of life remained unchanged by the centuries of chilly weather.

How slowly the time passes here, encompassed
as I am by frost and snow!

Frankenstein, Mary Shelley

Did you know? Mary Shelley's *Frankenstein* may have been inspired by the Little Ice Age. The year 1816 was so cold it was dubbed 'the year without a summer'. Young Mary was holidaying with friends in Switzerland; the weather was so bad she was encouraged by one of her friends, Lord Byron, to come up with an idea for a ghostly tale. It was there, huddled around the fire, that the idea for 'Frankenstein' first emerged – a novel set in, among other places, the North Pole.

SNOW in LIVING MEMORY

Just after the Second World War, Britain was victorious but exhausted. Little did it know it was about to face another battle. During the middle of December 1946, temperatures suddenly plummeted, reaching a low of –14°C (7°F). Heavy snow began to fall and, despite a brief respite over Christmas, carried on with such a force that 3 metre (10 ft) snow drifts brought the country to a standstill.

Roads were blocked, trains stopped, farm animals froze or starved to death and, at one point, London had only six days of coal left to burn before it ran out. The ground was frozen so hard, farmers couldn't lift their root vegetables; British potatoes – which had never been restricted, even in wartime – were rationed for the first time.

But the winter of 1946/47 wasn't the coldest Britain experienced. During the Big Freeze of 1963, a cold spell gripped the country with temperatures dropping to a record -22°C (-8°F). Snow lay on the ground for 62 consecutive days, rivers and lakes froze over and rural communities were cut off for days. Not even London escaped the chill – photographs from the time show people walking and cycling on frozen sections of the River Thames, milkmen on skis and even people ice skating in front of Buckingham Palace.

More recently, in March 1993, America experienced one of the most devastating snow events in living memory. The 'Storm of the Century' paralysed a third of the country as it bulldozed its way up the Atlantic coast over the space of three days, bringing huge snow drifts, coastal floods, blizzards, tornados and crippling temperatures. At its height, the storm stretched from Honduras to Canada, and wrought havoc over 26 states. The almost-biblical combination of snow, thunderstorms, winds and flooding caused billions of dollars of damage, killed more than 300 people and affected nearly half the population of the country. In places, snowfall was as high as 1.8 metres (6 ft), wind speeds nearly touched 150 miles per hour (241 kmh) and thermometers recorded lows of -24°C (-11°F). Incredibly, only three years later, as the eastern seaboard had just finished dusting itself off from the Storm of the

Century, another catastrophic snow event hit. The Blizzard of 1996 hit New York and the rest of the East Coast like a train, killing over 150 people and forced the federal government to shut down for nearly a week.

Will we see the Thames freeze again?

Even though parts of the Thames froze during the winter of 1963, it's unlikely that we'll see it happen again any time soon. This is for three reasons: first, the average winter temperature is too high – in 1814, the average January temperature was nearly -3°C (27°F). Now it's 1.4°C (34.5°F); the Thames also allows in more sea water than it used to – saltier water has a lower freezing point; the Thames also flows faster than it used to, also making it less likely to freeze.

Will we have another Ice Age?

According to climate models, Earth is due another Ice Age 50,000 years from now. However, new research at the Potsdam Institute for Climate Impact Research, Germany, suggests that man-made global warming will delay this event by an extra 50,000 years, meaning that vast ice sheets aren't now predicted to return to continental Europe and North America for 100,000 years, if at all.

THE ETYMOLOGY OF SNOW

While it's impossible to know exactly when the word 'snow' was first spoken, linguists can take an educated guess at its origins and follow the progression of the word as it spread across the globe. The theory goes that the root of the word comes from a language called 'Proto-Indo-European' or PIE, which was spoken by a group of people who lived roughly 6,500–4,500 years ago. Many modern languages descend from this eastern European single tongue – from Spanish to German, French to Italian, Urdu to Yiddish.

The PIE word for 'snow' is written as $sneig^{wh}$ (which may have been pronounced something like 'sney-gwah'). From that root, the word has stayed remarkably similar over vast geographical areas and across a huge expanse of time – from the Old English *snāw*, which then became '*snow*', to the Icelandic '*snjór*', Swedish *snö*, Irish *sneachta*, Russian *sneg*, Latvian *sniegs*, German *Schnee*, Dutch *sneeuw,* Ancient Greek *nípha* and Sanskrit *snēha* all these versions share a striking similarity. Interestingly, other countries who use words beginning with 'n' for snow – such as France's *neige* – may seem as if they don't belong. And yet, take away the 's' from the original PIE root and we can see the resemblance. Suddenly, the Welsh *nyf,* the Italian *neve*, French *neige* and Spanish *nieve* become part of the family.

Did you know? *Nivisphobia* is the fear of avalanches. Fear of snow is *chionophobia* and *pagophobia* is the fear of ice or frost.

SNOW DIALECT

Language has a wonderful way of evolving and becoming nuanced. Snow has so many different forms – it changes and mutates as it falls, lands and melts – that it's no surprise different regions have come up with their own words to describe what they experience. From the subtle differences between different types of snowfall to trying to describe the worst a snow storm can throw at you, here are just some of the more unusual or long-forgotten:

BLENKY – an old West Country word which describes very light snowfall, barely a smattering. 'Blenks' was an old word for ashes – the idea being that the snow looked like the white flecks that wafted down from a chimney or bonfire. The Scots would have called this kind of light snow shower 'flindrikin', a word that also means flimsy or frivolous.

BLIND SMUIR – a fantastic historic Scottish word for a snow drift. 'Smuir' meant to 'smother' or 'suffocate', so a 'blind smuir' was a snow storm that not only blinded you, but also choked.

> *That dolefu' day, in whilk the lift*
> *Sent down sic show'rs of snaw and drift,*
> *To smuir his sheep – he was sae glift*
> *He ran wi' speed*
> *To save their lives – ah! dreadfu' shift*
> *It was his dead.*

'Poems', Berwickshire Sandy

ONDING – An 18th-century word, originally from the Middle English *dingen*, which means to hit repeatedly, 'onding' is heavy, unrelenting snow or rain.

> It was a very grey day; a most opaque sky,
> 'onding on snaw', canopied all; thence flakes
> fell at intervals, which settled on the hard
> path and on the hoary lea without melting.

> Jane Eyre, Charlotte Brontë

SNOW-BROTH – a medieval phrase meaning melted snow or slush, it even turns up in Shakespeare:

> ... Lord Angelo; a man whose blood
> Is very snow-broth; one who never feels
> The wanton stings and motions of the sense
> But doth rebate and blunt his natural edge

> Measure for Measure, William Shakespeare

POUDRE – the French Canadians used 'poudre' to describe powdery snow (from the Old French *poudre* meaning powder or dust) until the early

20th century. Interestingly the Scots use a similar word, *'snaw-pouther'*.

> *A 'poudre' day, with its steely air and fatal frost, was an ill thing in the world; but these entangling blasts, these wild curtains of snow, were desolating even unto death.*

> *Pierre and His People*, Gilbert Parker

ICE-SHOGGLES – old Yorkshire dialect for 'icicles'. Other regional gems from across the UK include clinker-bells, daglers, ice-lick, izles, snipes and tanklets. Interestingly, many words in Yorkshire dialect have Viking origins – the word 'glocken', which describes the point at which snow begins to thaw, comes from the Icelandic *glöggur*, which means 'to make clear'.

. . . thou knowest, winter tames man,
woman, and beast.

The Taming of the Shrew,
William Shakespeare

LIVING WITH SNOW

SNOW COMMUNITIES

People who live outside the Arctic imagine it to be a barren and hostile place. And yet, for the dozens of indigenous groups – some of whom have lived there for over 20,000 years – the Arctic is a bountiful, rich environment capable of sustaining life all year round.

What we call the Arctic is actually a region that encompasses all, or parts of, eight different countries: Norway, Sweden, Finland, Denmark, Iceland, Canada, Russia and the United States. Indigenous groups make up about 10 per cent of the 5 million people who live in the Arctic. The Antarctic, by contrast, is the only continent never to have had its own indigenous population.

To survive long, harsh winters and brief summers that characterise the region, the people of the Arctic have had to adapt to radically shifting weather patterns and migrating wildlife; some groups settled along coastlines, skilfully exploiting fish stocks and marine mammals, while other native Arctic populations adapted to a semi-nomadic life in the interior, following migrating herds, hunting, trapping and foraging.

The Aleut, for example, who originally inhabited the Alaskan Aleutian Islands and the Kamchatka area of Russia, developed a subsistence lifestyle closely tied to the sea and coastline, fishing and hunting sea lions and seals, while the Athabaskan, a group that spread across Alaska, Yukon and Northwest Territories, thrived by migrating seasonally, travelling in small groups to fish and hunt caribou, moose, beaver and rabbit.

The Inuit, who make up almost 90 per cent of the population of Greenland, and live in parts of Canada and Alaska, were traditionally skilled fishers and hunters – relying on whales, caribou, walruses, polar bears, seals and other Arctic prey – while the Saami people, who cover a vast region that touches parts of Sweden, Norway, Finland and Russia, developed a way of life that not only focused on semi-nomadic reindeer herding but also sheep herding, fishing and fur trapping.

Many of the 500,000 indigenous people who call the Arctic region home still live their lives through subsistence herding, hunting and fishing, and keep their cultural and language traditions alive. One of the greatest challenges that many of these communities will face over the coming years will be the effect of climate change on their traditional way of life (see **CLIMATE CHANGE** page 62). Scientists have predicted, for example, that entire coastal communities may be forced to relocate as the permafrost melts and coastlines erode without the buffering effects of sea ice. The rapid pace of warming in the Arctic also alters the balance of wildlife, putting the future of many indigenous people – who rely on Arctic species for both food and other essentials – at risk.

SNOW HOMES

Insulation materials are measured in R-values. The R stands for 'thermal resistance' and basically measures how easy it is for heat energy to travel from one side of a material to the other. The higher the R-value, the more resistant and, therefore, insulating a material is.

So, when you're building a shelter, it's interesting to know what the R-value of a material is. A typical sheet of plasterboard has an R-value of about 0.5; a brick about 0.8. But did you know that 2.5 centimetres (1 in) of snow has an R-value of 1? That means if you build a shelter with walls of snow 25 centimetres (10 in) thick it would have the same R-value as 15cm (6 in) of fibre insulation.

People who live with, and understand the thermal properties of, snow have exploited this fact. The Inuits of Northern Canada and Greenland, for example, are famous for building **IGLOOS**, temporary shelters traditionally built on winter hunting trips. The igloo is a masterclass in resourceful, low-impact architecture – the structure is created from blocks of compacted snow, which are cut and stacked in a continuous spiral to form a dome-like dwelling. The air pockets trapped within the snow act as an excellent thermal insulator, creating a snug environment inside, even as the outside temperatures plummet (see **MAKING AN IGLOO** page 144)

Many of the igloo's design features aid in the process of keeping warm and safe – its inherent structure is so strong it can take the huge weight of extra snow piled on top, while the heat created by body temperature inside the igloo causes the surface of the blocks to melt slightly and refreeze, sealing up any draughty gaps. The short entrance tunnel reduces wind flow into the igloo, while skins and furs can add extra layers of comfort to the interior. Inside, heat rises, so sleeping platforms are built up from the floor, keeping the inhabitants in the warmest part of the igloo. Extra snow is often banked up against the side

◇◇◇

Did you know? The 16th-century explorer, privateer and salty seadog Martin Frobisher is thought to be the first European to clap eyes on an igloo, during a visit to Baffin Island in 1576. After an initially warm reception from the indigenous Inuit, five of ship's crew disappeared after rowing to shore, followed by an arrow in the bottom for Frobisher. In revenge, Frobisher captured an Inuit man and brought him back to England where, in an ironic and cruel twist of fate, he died of a cold.

◇◇◇

of the igloo, to add another layer of insulation. And then, come the summer, when the igloo is no longer needed, it simply melts away without leaving a trace.

Other indigenous cultures have also made use of snow homes. The Quinzhee is another kind of temporary structure, often used in survival situations. The word comes from the Athabaskan language (see **SNOW COMMUNITIES** page 91), spoken by the Slavey and Sahtú people of Canada, and describes a snow shelter made from a large pile of loose snow, shaped and then hollowed out to leave thick, insulating walls. To build a quinzhee, snow is piled into a mound roughly 2 metres high by 4 metres wide (6½ by 13 ft). The pile is left to settle for at least a few hours, a process called 'sintering' whereby snow begins to bind together under compression. The inside is then scooped out, leaving a space big enough to sit or crouch in. The Finnish have a word for a similar shelter, called a **LUMITALO**.

People who explore snow-covered regions are taught the art of making temporary shelters for emergency situations. The simplest of makeshift shelters is a **SNOW TRENCH** or snow grave; a 1 metre (3 ft) deep trench big enough for a person to lie down in, which is covered with a tarp or sheet held down with snow blocks or any heavy pieces of equipment to hand.

While snow trenches and quinzhees work well on flat ground, mountaineers have long relied on **SNOW CAVES** to provide emergency respite on long ascents. To build a snow cave you need a steep slope. The idea is to tunnel horizontally into the slope, and then start to dig upwards, creating a large chamber that is above the original entrance tunnel. With the main chamber dug out, snow can be used to temporarily block up the entrance tunnel, effectively sealing yourself into a heat-trapped space. A small ventilation hole is poked through the wall of the chamber, out into fresh air.

Did you know? The largest igloo ever made was built by a Swiss team of 18 people in Zermatt, Switzerland. It was 10.5 metres (35½ ft) high and, internally, its diameter reached 12.9 metres (42 ft). (To give you a comparison, a London double-decker bus is only 4.4 metres /14½ ft) high by 10 metres/33 ft long.)

SNOW CLOTHING

Modern snow clothes combine high-tech fabrics, lightweight construction, and life-saving insulation, but humankind has long had to dress against the elements. The **OLDEST WINTER HAT** ever found belonged to Otzi, a Bronze Age man found frozen in the glacial mountains that straddle Italy and Austria. His 5,300-year-old body, along with much of his clothing and equipment, were preserved in the ice and give us a glimpse of how prehistoric people would have clothed themselves to survive the cold. Otzi's hat was made not from wool but bearskin, and fitted closely to his head, rather like a modern bobble hat or beanie. Pieces of bearskin had been stitched together to form a bowl-shape and the hat was kept in place with a narrow chin-strap. The precursors to the modern knitted ski bobble hat, however, can be seen in the *chullos* of the Andean people. These woollen, thick hats with ear flaps insulate the wearer from the extremes of mountain weather and have been used by indigenous communities at least as far back as the Incan Empire.

The earliest examples of SNOW GOGGLES come from the Inuit and Yupik people of the Arctic (see SNOW COMMUNITIES page 91). Unlike modern ski goggles, which use clear plastic to shield the eyes, these ingenious pieces of eyewear were traditionally made from driftwood, bone or antler into which a long, narrow, horizontal slit had been carved. The snow goggles were tailored to fit tightly the individual wearer's face so that the only light that you could see came through the slit. This had three benefits: not only did it stop the wearer getting snow or sea spray in his eyes, but it also reduced glare and minimised the chance of snow blindness. The design of the goggles also improved the wearer's visual acuity: the narrow slit provided the same optical effect as a pinhole lens, in that the smaller the aperture, the sharper the image and the greater the depth of field.

When modern humans migrated out of Africa into Glacial Europe about 45,000 years ago, there is no doubt that they would have needed highly insulating, specialised clothing to cope with the cold climate they encountered. We don't know for sure what these early pioneers would have worn but, lacking access to wool or synthetic fibres, they would have been forced to rely on animal skins and fur. There's even a theory that 125,000 years ago Neanderthals, who adapted to live in frozen climates, would have needed to cover up all but a small portion of their bodies to survive the extremes of Ice Age Northern Europe. We'll never know who invented the first warm winter coat – ivory figurines from Russia that date back 24,000 years show what look like people wearing

---

Did you know? The word 'parka', far from being a 1960s Mod term, actually comes from the Nenets people, one of the many indigenous communities that make the Arctic their home. The word was first recorded in 1625 by travel writer Samual Purchas in his book *Purchas His Pilgrimes* (Volume III):

Now for the manner of the Samoits in the Journey, their upper Coat is called a Parka, which is for the most part of Deere-skin, and some of white Foxe or Wolverin, which they weare the hayre or furre outward.

---

the first parkas – and the Inuit have long worn this type of jacket made from animal skin and trimmed with fur. The *'amauti'* is the name for the parka worn by Inuit women, which has a built-in baby pouch nestled below the hood, and is fastened at the waist with a tie to prevent the young child from slipping down.

SNOW EQUIPMENT

For thousands of years, people have attempted to craft specialised equipment that not only helps them survive snowy conditions but actively utilises the physical properties of snow to make travel quicker and transportation easier. Take **SNOW SHOES**, for example. Before the physics of weight distribution were fully understood, people already intuitively knew that, to walk across deep snow, it was important to create as large a footprint as possible. Perhaps taking their cues from the natural world – and looking at the design of animal paws – as early as 6,000 years ago, the snow shoe was already fully developed and not unlike modern equivalents. The oldest physical example found to date came from the melting snow of a glacier in the Italian Dolomites; the shoe was made from a 1.5 metre (5 ft) length of birch, bent into an oval shape and lashed together with twine – carbon dating put the shoe around 4000 BC.

Both China and Scandinavia have a long history of skiing. Both claim to be the first region to strap on a pair of SKIS but the evidence has yet to provide a definitive answer. What is agreed, however, is that the first skiers were hunters, not pleasure seekers. Some of the earliest clues come from petroglyphs, pictures carved into the walls of ancient caves. In the Altai Mountains, a place where China, Mongolia, Kazakhstan and Russia converge, there's an image of a skier hunting an ibex scratched into the rock – archaeologists can't agree on its age, putting it anything between

3,000 and 10,000 years old. At the same time, petroglyphs of skiers have turned up across Norway, Sweden, Finland and Russia, the oldest to date being 5,500 years old.

Archaeologists are on firmer ground when it comes to ancient remains of skis – the oldest, an 8,000-year-old tip of a wooden ski that turned up in peatland near Lake Sindor, Russia. What's interesting about many of the early finds and depictions of skis is that the skier is often shown using just one **SKI POLE**. Researchers think that early skiers would have used the pole – which was shaped like a paddle, rather than a stick – as a kind of steering rudder when they were hurtling down a mountain at speed. The standing position was also different – leaning backward, rather than forward – and the shovel-shaped pole would have also been used as a spade for digging snow, a scoop for clearing ice and a weapon for hunting. Interestingly, people in the remote, snowy Altai Mountains still ski – on homemade wooden skis covered with horse-hide – using this one-pole technique.

There's even more disagreement when it comes to the origins of the **SNOWBOARD**. Often seen by many as a relative newcomer, there are some tantalising clues that it may be a much older piece of snow equipment than first thought. The conventional story is that snowboarding really began in the 1960s, when Sherman Poppen created the 'Snurfer' by fixing two skis together and attaching a rope to the front. The Snurfer – a marriage of the words 'snow' and 'surfer' – was a huge hit, kick-starting a craze and inspiring further developments and refinements that culminated in the modern snowboard. The idea of standing up on one wide ski, rather than two, was certainly nothing new, however. People in the Turkish Kaçkar Mountains have been riding sideways on a flat, snowboard-like device called a *lazboard* for at least four centuries, the only difference being that the rider holds onto a rope at the front for balance and a stick at the rear for steering. There are also tales of Austrian miners as far back as the 1500s using, and racing on, wooden snowboards they called *Knappeross*, and a Swiss version called a *Rittpratt* appearing in the Alps at a similar time.

As with skis, the earliest **ICE SKATES** were tools for survival not fun, according to archaeologists. Skates made from animal bones have been discovered across Scandinavia, some dating back as far as 5,000 years. Made from shaped cow or horse bones (usually the shin bone), the skates would have been strapped to the feet with leather bindings, and were invented to make it easier to cross the many frozen lakes that characterise those parts of the world, especially Finland. And, as with skiing, the technique would have been different too: research suggests that these early skaters punted themselves along with long sticks, which they straddled, rather than pushing off from each foot. Amazingly, this design of ice skate persisted into medieval times; the Museum of London, for example, has a pair of 12th-century bone ice skates on display, while the writings of medieval chronicler William Fitz-Stephen describe Londoners using poles not only to push themselves along on their skates but also for 'ice-jousting', a game where players skated towards each other with poles outstretched with the aim of knocking each other over.

What's clear is that, in regions with consistent snowfall, disparate communities have come up with surprisingly similar equipment. Take the sled or **SLEDGE**, for example. Across the globe, archaeology has uncovered similar objects, created by people separated by vast distances of time and geography. Wooden sledges – a platform with smooth wooden runners very like the kind that children still use today – have been discovered in Viking burials, pictured on Ancient Egyptian tombs (to use on damp sand rather than snow), and are still used by traditional Inuit groups today; the *qamutiik* – the Inuit sled – would have traditionally been made from wood, a material scarce in the Arctic. While coastal groups could use driftwood that floated onshore, many of the inland Inuits had to travel huge distances to obtain the timber needed to make the runners. Other materials were also used if wood wasn't available, including whale or walrus bone, walrus ivory, antlers or even frozen animal skins.

◇◇

Did you know? The words 'sled', 'sledge', 'slide' and 'sleigh' are all related and come from the ancient Proto-Indo-European root *sleidh-* meaning to 'slip' (also see THE ETYMOLOGY OF SNOW, page 82).

◇◇

I wonder if the snow loves the trees and fields, that it kisses them so gently? And then it covers them up snug, you know, with a white quilt; and perhaps it says, 'Go to sleep, darlings, till the summer comes again.'

Through the Looking Glass, Lewis Carroll

CELEBRATING
SNOW

SNOW FESTIVALS

Winter solstice lands at the end of December and marks the shortest day, an event that's been celebrated for thousands of years. Officially the first day of winter, it actually represents the critical tipping point in the year when we can begin to look forward to longer, warmer days and the return of the sun.

Many cultures have historically marked this important time: in China, the Dongzhi Festival traditionally involved families coming together to enjoy a specially prepared meal, visiting ancestral tombs and celebrating the dead, while in Ancient Rome, the winter solstice was honoured at the Feast of Saturnalia, a seven-day binge of gift-giving, sacrifice, misrule and partying.

During this time groups would also dress up in masks and costumes, singing, acting and travelling from house to house. These 'mummers', who derived their name from the Greek god of satire Momus, are thought to be the early precursors to modern-day carol singers.

Before the advent of Christianity, Scandinavian communities enjoyed 'Juul' or 'Yule', twelve days of great fire-lighting to symbolise the blaze of the returning sun (and the origin of the 'Yule log'), while in Iran, families celebrate Yalda Night at the winter solstice, staying up well after midnight to eat, drink, tell stories and read traditional poetry. Pomegranates, watermelons, apples and other red foods are eaten to symbolise life and the returning sun.

It's interesting that, traditionally, festivals don't celebrate snow or cold weather, but instead look to banish winter and encourage the return of spring. Culturally, this makes sense, as for many marginal and subsistence communities, winter was a season of hardship and endurance. The London Frost Fairs of the Little Ice Age (see page 74) between the 17th and 19th centuries certainly made the most of an extended period of freezing weather but, at their heart, were an excuse for revelry rather than a celebration of snow as an integral part of the human experience. Writer John Evelyn described one fair, when the Thames froze for an entire two months, at the end of the 1600s:

> Coaches plied from Westminster to the Temple, and from several other stairs too and fro, as in the streets; sleds, sliding with skeetes [skates] a bull-baiting, horse and coach races, puppet plays and interludes, cooks, tipling and other lewd places, so that it seemed to be a bacchanalian triumph, or carnival on the water.

And yet, in more recent years, many snow-bound communities have initiated festivals that explicitly celebrate snow and ice and, in particular, its artistic potential. The first ice palace, for example, is thought to have been built in the winter of 1739 in St Petersburg, for Empress Anna Ivanovna, to celebrate Russia's victory over the Ottoman Empire, and came complete with an ice garden filled with carved frozen trees, birds and an elephant. The city still builds one every year. Since then, many places have embraced ice and snow festivals as an opportunity for creative expression and tourism, including St Paul in Minnesota, Quebec City, Saranac Lake in New York, Sweden's 'Snöfestivalen' in Kiruna and, the largest of them all, Harbin International Ice and Snow Sculpture Festival, in Heilongjiang, China.

◇◇◇

Did you know? Solstice comes from the Latin word *solstitium*, meaning 'stationary sun'.

◇◇◇

SNOW AND CHRISTMAS

Dashing through the snow
On a one horse open sleigh
O'er the fields we go,
Laughing all the way
Bells on bob tail ring,
making spirits bright
What fun it is to laugh and sing
A sleighing song tonight

'Jingle Bells', James Pierpont

What are the odds of a White Christmas?
It depends what you mean by a 'White Christmas'.
In the UK, the Met Office needed just one snowflake to fall on Christmas Day for it to qualify.
Using that definition, 38 of the past 54 years have
seen snow fall somewhere across Britain – odds of
about 70 per cent. Most of us, however, imagine

a White Christmas to involve a more generous, quilted blanket of snow. Over a similar period, there's only been a significant covering of snow (where more than 40 per cent of the weather stations in the UK reported ground cover) four times – slashing the odds of a truly White Christmas to a rather less jolly 8 per cent.

In the US, a 'White Christmas' is defined as having at least 2.5 centimetres (1 in) of snow on the ground. The huge geography of the country makes national probabilities difficult to calculate – ranging from a 90 per cent chance of snow in Anchorage, Alaska to just 1 per cent in Florida. That said, one US hydrologist – Ethan Gutmann from the National Center for Atmospheric Research – analysed the data from 1,000 weather stations across the country and came up with a current average probability of 23 per cent. Interestingly, he also looked at the readings around the 1940s – when Bing Crosby's version of Irving Berlin's festive song first came out – and found that the probability of a White Christmas during this period was much higher, at 33 per cent.

Why do we associate Christmas with snow? If the chances of snow are slim for most of us during the festive season, why do we associate it so closely with Christmas? One theory puts the blame firmly at the feet of Charles Dickens, whose famous novella, *A Christmas Carol*, set the story of redemption and second chances against a backdrop of an often idealised, snow-covered London:

> *... and they stood in the city streets on Christmas morning, where (for the weather was severe) the people made a rough, but brisk and not unpleasant kind of music in scraping the snow from the pavement in front of their dwellings, and from the tops of their houses, whence it was mad delight to*

the boys to see it come plumping down into the road below, and splitting into artificial little snow-storms.

The book was published in 1843 and is credited with creating the iconic image of the white-blanketed Christmas, along with other writers and artists of the time such as Pieter Bruegel and Abraham Hondius. Across the water in America, the vastly popular poem 'The Night Before Christmas' by Clement Clarke Moore also painted a picture filled with flourishes of a rosy-cheeked Santa, gifts and a silent, snow-covered night:

The moon on the breast of the new-fallen
    snow,
Gave a lustre of midday to objects below,
When what to my wondering eyes did
    appear,
But a miniature sleigh and eight tiny
    rein-deer

It's interesting that many of the artistic works of the Victorian period were created by people who had experienced the coldest decade since the 1690s, between 1810 and 1820 (including London's last Frost Fair in the winter of 1813). It's thought that, for example, of Dickens's first nine Christmases as a child, six were white.

In the same year as Dickens published *A Christmas Carol*, inventor and civil servant Henry Cole sent the first Christmas card. The idea was a hit and the Victorians sent and collected Christmas cards in their thousands; the cards often showed an idealised version of Christmas and helped to cement many of the familiar images we associate with the festive season, including red-breasted robins, Christmas trees, crackers and, of course, snow-covered landscapes.

SNOW and FOLKLORE

Advice is like the snow. The softer it
falls, the longer it dwells upon and
the deeper it sinks into the mind.

'Confessions of an Inquirer,'
Samuel Taylor Coleridge

Before the days of reliable weather forecasting, people relied on observations of the natural world and ancient superstitions to predict the arrival of snow or a winter storm. These seemingly irrational beliefs would not only give comfort and reassurance during times of physical hardship but were also a way of passing on oral traditions and folklore in a society where few could read or write. Farmers, rural workers and seafarers in particular had numerous sayings not only to help them predict a cruel winter but also to forecast the welcome return of spring:

MANY HAWS, COLD TOES – This old Yorkshire saying has numerous regional equivalents and, warns that a plentiful crop of hedgerow berries in autumn (such as rosehips, blackberries and hawthorn berries) denotes a hard, snowy winter to come. The superstition has its basis in sound reasoning – that nature is providing extra supplies to help the birds through the coming season – although science has yet to prove the theory.

THE CAT TURNING ITS BACK – Another belief, widely recorded from the mid 1700s, but undoubtedly in use earlier, claims that if a cat sits with his or her back to the fire, it signals a hard frost is about to come.

CLEAR MOON, FROST SOON – A saying that correctly links the presence of a clear night sky with an increased likelihood of a frost. This is because, without the insulating presence of clouds, clear skies allow the Earth's heat to escape into the atmosphere, cooling the ground temperature.

IF THERE'S A HALO AROUND THE MOON, 'TWILL SNOW SOON – The theory that, if you see a halo or circle around the moon, snow will shortly follow, has some basis in truth. Halos are formed by the light from the sun or moon refracting as it passes through ice crystals in clouds very high in the sky. These clouds often precede the arrival of a low-pressure system, which may bring a snow storm (or rain).

PALE MOON DOTH RAIN, RED MOON DOTH BLOW, WHITE MOON DOTH NEITHER RAIN NOR SNOW – We know that it can't snow if there isn't dust in the air. The more dust particles that are in the air, the greater the chance that mois-ture will have something to cling onto and create droplets that fall either as rain or snow. When you look at the moon through dusty air, it can appear pale or a dusky red. If the air is less dusty, the moon looks whiter.

WHEN THE SNOW FALLS DRY IT MEANS TO LIE, BUT FLAKES LIGHT AND SOFT BRING RAIN OFT – Dry snow is generally produced in much colder temperatures – 0°C (32°F) or less – making it less likely to melt away. Wet snow, which falls through air that is slightly warmer – above freezing – makes lovely large, fluffy snowflakes.

EATING SNOW

If snow is free, clean and abundant, why don't we eat and drink more of it? While history gives us a few famous examples of snow recipes – Alexander the Great and Emperor Nero both enjoyed an ice slushy, apparently – there are few examples of snow being used in recipes or as a recommended source of drinking water.

Apart from the obvious slushy headaches, tooth sensitivity and chilling effect of eating mouthfuls of snow, it seems there are more pressing reasons not to eat too much of the white stuff.

Groundwater and surface water – the kinds of water we usually drink – are not 'pure' water; they also contain tiny amounts of dissolved substances such as minerals and organic matter that are thought to be beneficial to human health. Fresh snow and falling rain water are, on the other hand, demineralised. In other words, water that comes from new snow or raindrops is pure and doesn't contain any of these trace elements. According to the World Health Organisation, drinking water that contains no or very few essential minerals for any significant length of time has been linked to a number of health problems, including weakness, fatigue, bone decalcification and muscle cramps. Such problems have been reported in climbers, for example, who have relied for long periods on melted snow for drinking water, without adding the necessary minerals. That's not to say that, in emergencies, snow isn't an important tool for survival – experts often recommend that you warm it up first, however, as eating or drinking frozen water can lower the body's temperature to dangerously low levels.

Another interesting reason why it's not a great idea to eat snow in any large quantity is that, as snow sits on the ground, it slowly collects bacteria. One recent study tested how long it took freshly fallen snow to 'go off'. While new snow has virtually no bacteria, the experiment revealed that, after half a day, it becomes contaminated; levels of bacteria increase with pollution levels, however, so city snow goes off quicker than rural snow and the colder the weather, the slower the bacterial growth. Urban snow is also prone to collecting toxic particles, especially from exhaust fumes.

It's important not to be alarmist, however. Few

pleasures are greater than catching snowflakes on your tongue or eating the odd mouthful of nature's sorbet. The Yupik of Alaska, for example, famously make a version of ice cream called *akutaq* (which means 'mix together') which blends animal fats, berries and fresh snow into a whipped foam. Wartime recipes for pancakes often included a spoonful or two of snow to make them lighter, while 'maple taffy' is a well-known confectionary in parts of Canada, and made by pouring hot maple syrup onto snow. So, if you're going to eat snow, just follow a few simple rules: no old snow, no dirty snow, no ploughed snow, no city snow and, definitely, *definitely*, no yellow snow.

◇◇◇◇◇◇◇◇◇◇◇◇◇◇◇◇◇◇◇◇◇◇◇◇◇◇◇◇◇◇◇◇◇◇◇◇

Did you know? In Ancient Greece, it was thought that drinking water from melted snow would give you 'goitres', a thyroid disorder which causes your neck to swell. Although they didn't know why, they weren't too far off the mark – one of the causes of goitres is lack of iodine and, as we know, snow water – unlike groundwater – doesn't contain any minerals, including iodine.

◇◇◇◇◇◇◇◇◇◇◇◇◇◇◇◇◇◇◇◇◇◇◇◇◇◇◇◇◇◇◇◇◇◇◇◇

In seed time learn, in harvest teach,
in winter enjoy.

'Proverbs of Hell', William Blake

SNOW
PLAY

SNOWMEN

Humans are endlessly creative. It's easy to imagine that, as soon as early humans saw and experienced snow, they began to explore its creative potential. We're imitators at heart – we re-create what we see – so it's not difficult to picture that somewhere, in the long distant past, our ancient ancestors rolled snowballs, sculpted with ice and packed snow into human shapes.

And yet evidence for early snowmen is rare. The earliest drawing we have of a snowman is tiny, tucked in the footnotes of a medieval illuminated manuscript called the *Book of Hours* from 1380. It shows what historians believe to be an unhappy, odd-shaped snowman, charring his bottom by an open fire. And yet for the first picture of what we would recognise as a traditional snowman we have to wait until 1603 and a book called the *Petits Voyages*, an atlas and book of engravings which include one of explorer Willem Barentz talking to Inuits; in the background, small, but clear as day, is a snowman.

There have been some famous snowmen too. The 16th-century painter and art historian Giorgio Vasari described one made by Michelangelo at the request of the Medici family:

> ... *one winter, when a great deal of snow fell in Florence, [Piero de' Medici] had him make in his courtyard a statue of snow, which was very beautiful.*

A snowman was also one of the first things ever photographed – in 1845 photography pioneer Mary Dillwyn snapped a picture of a man and woman building a snowman. Not only is it one of the earliest photographs ever taken, but it's one of the first to take the art form outside the studio and show the real, spontaneous moments that make up family life.

The world's biggest snowman ever built was actually a snow woman. The residents of Bethel in Maine, USA, helped by people from the surrounding towns, used nearly 6 million kilograms (6,000 tons) of snow to sculpt a snow woman who measured 37.21 metres (122 ft) high (only a few metres shorter than the Statue of Liberty from base to torch). It took a month to complete, had arms made from two 9 metre (30 ft) spruce trees and eyelashes made from eight pairs of skis.

The smallest snowman ever made measured just 0.01 mm across (about a fifth of a human hair). Not actually built from snow, it was made by scientists at the National Physical Laboratory in the UK with two tiny balls usually used for calibrating electron microscopes. A minute ion beam was then focused on the snowman to etch two eyes, a nose and a little smile.

HOW TO BUILD A SNOWMAN

We all know how to build the basics of a snowman – it's child's play. But here are a few tips to help get the ball rolling...

The white stuff

You need a certain type of snow for both snowmen and snowballs. The dry, powdery stuff that falls when it's really cold just won't stick together, so you need the nice, slightly wet snow that tumbles in large, fluffy flakes around temperatures just a degree or so above freezing (see WET SNOW page 22). A quick test – press the snow together in your hands; if it stays in a compact ball that can be thrown up in the air and caught, you're good to go.

The surface matters

Choose the right ground surface for your snowman. Some common sense here – build your snowman in a shaded spot, away from the melting effect of direct sunlight. Pick a relatively flat surface but avoid asphalt, as it absorbs more heat from the sun than other hard surfaces or grass, causing snow and ice to melt more quickly.

Start off by hand

Don't try to roll a snowball straight away. You need to pack a large snowball by hand first – try and get it to about a ruler's width (30 centimetres/12 in), densely packed together, before you put it on the ground and start rolling.

Work out your end point

Decide where you want your snowman to sit and have enough space to roll the snowball so you end up in the right place. Start the ball rolling. You'll notice that you're creating a cylinder not a sphere, so you'll need to keep changing direction or turning the snowball through 90°.

Build a large, solid base

The first snowball is the base of your snowman. Once you've positioned it, flatten the top of the snowball ready for the body.

The ideal snowman

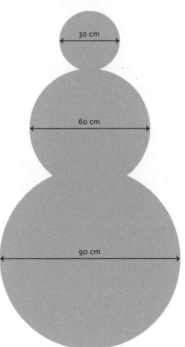

30 cm

60 cm

90 cm

Build a medium body

Again, flatten the bottom of the body, so the two sections make good contact. Top tip – if you're struggling to lift the body on top of the base, prop a plank against the base and roll the body up into place. Now, flatten the top of the body, ready for the head.

Attach a small head

The optimum snowman ratio of base:body:head is 3:2:1 from bottom to top. The base needs to be strong enough to support the combined weight of the body and head, but after a certain size snow-balls can lose stability as you can't apply enough pressure to pack the snow tight enough. For added stability you can pack extra snow around the joints where each section meets.

SNOWBALLS

The same kind of snow that's good for snowmen is, unsurprisingly, good for snowballs (see **HOW TO BUILD A SNOWMAN** page 136) – not too slushy, not too powdery. There's a bit more room for manoeuvre with snowballs, however, as the heat of your hands can help melt the ice crystals and stick them together. While hard-core snowballers live by the rule 'Bare hands, better snowballs', for those of us who enjoy having a full complement of fingers and thumbs, knitted gloves are better for snowball fights than mittens. This is because you can exert more packing pressure with your fingers kept separate and also gloves tend to lose more heat than mittens. You can also grip the snowball more tightly with a glove, allowing for a more accurate throw.

Making a snowball

For the best technique for snowball formation, don't try to create your snowball in one squeeze. Gentle pressure and constant turning of the snowball (not unlike making meatballs) gives you a rounder, more compact shape and lessens the risk of the thing exploding in your hands if you squeeze too hard.

Connoisseurs might finish off the process with a quick smoothing down of the surface to create the perfect projectile or, if they're in the mood for a dirty fight, dip their snowballs in water to create lethal ice balls called 'soakers'. Not recommended if you want to play fair.

In Japan, snowball fighting has been elevated

to a professional sport called *Yukigassen*. Two teams, each with seven members, have 90 ready-made snowballs each and must battle it out to win the tournament. With rules not dissimilar to dodgeball, each team tries to eliminate players by striking them with a snowball. Whoever is left with more players on the field when the whistle goes, or captures the other team's flag, wins.

Throwing tips

If you're not having much luck hitting your targets, take a few tips from baseball pitchers – the masters of accurate, quick throwing. There are dozens of coaching videos online but, in essence, you need to stand with your feet shoulder width apart. Turn your body so that you are side on to your target, rather like an archer with a bow, and grip the ball with your fingers not your palm. At the start of the throw hold both hands in front of your chest, like a pitcher, and when you throw turn your chest to face your target. As you release the snowball from your hand, point your fingers towards your target – it will improve your accuracy.

◇◇

Did you know? The largest snowball fight was enjoyed by 7,681 people at an event in Saskatoon, Saskatchewan, Canada, in 2016. A 2017 attempt to break this record, in Ocean County, New Jersey – which planned to include 9,000 participants – had to be cancelled because of too much snow.

◇◇

When I'm playing out in the snow, why can I see my breath?

When you breathe, a number of things come out of your mouth. Along with carbon dioxide, nitrogen and oxygen, when you exhale your breath also contains about 5 per cent water vapour. When it's very cold outside, this water vapour cools very quickly and condenses into tiny droplets of water or ice that you can see.

MAKING AN IGLOO

While the Inuit have igloo construction down to a fine art, it's not as easy as it looks. Here's how to make a simple, pared-down version for your backyard – just for fun, not for survival. You need *two* people – one working from inside the igloo, and one to pass over the blocks

Choose the right snow

Powdery snow doesn't make a good building material – you'll need deep, compacted snow to create blocks. If you struggle to find suitably compact snow, you have two options – one is to pile the snow into a huge mound and leave it overnight to bind together – a process called sintering – (see 'Quinzhees' in **SNOW HOMES**, page 94) or you can trample the snow underfoot.

Create a circle

An igloo that's 2.5 metres (8 ft) or less across is safest if you're a beginner – use a string and peg to mark out a perfect circle of the right diameter.

Cut out your blocks

Use a bread knife or handsaw to cut out rectangular blocks. The snow will dictate how large the blocks can be – traditional Inuit igloos are made from huge blocks about 90 centimetres long x 40 centimetres high x 20 centimetres thick (35 x 16 x 8 in) – but local snowfall will probably only allow for smaller blocks. Lay one complete circle of blocks. That's your foundation level.

Form the curve

Using your knife or saw, cut a gentle slope around a third of the circle as in the picture overleaf. This allows you to start building the rest of the blocks in an upwards spiral. If you are right-handed you'll find it easier to build in an anti-clockwise direction, and vice versa for left-handers. Work from the inside of the igloo; get the other person to pass you the blocks over the wall.

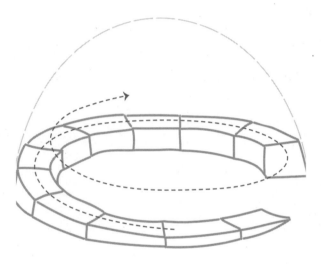

Shape the blocks

Using your saw or knife, shave off the bottom of each block so it tips slightly forward when it's positioned on the block below. This is how you create the dome shape. You can also tweak the shapes of the blocks with the knife or saw as you push them together, making as neat a fit as possible.

◇◇◇

Did you know? Traditional Inuit igloo builders would craft different shelters depending on their purpose. One man, working alone, could throw up a snow dome designed to be an overnight shelter in around an hour. Larger, more permanent igloos – created for family living – could take two days.

◇◇◇

Make a hole

For safety, make the blocks slightly smaller as you get nearer the top. At this point, break off and cut an entrance hole – as small as possible – through the wall. The traditional Inuit way is to do this from the inside, after the igloo is complete, but it's safer not to block yourself in.

Finish off

Close the final gap in the roof with a block cut to shape. Pack snow into any gaps between the blocks, both inside and out, to create a smooth, stable surface.